# ALPHONSE IN AUSTIN

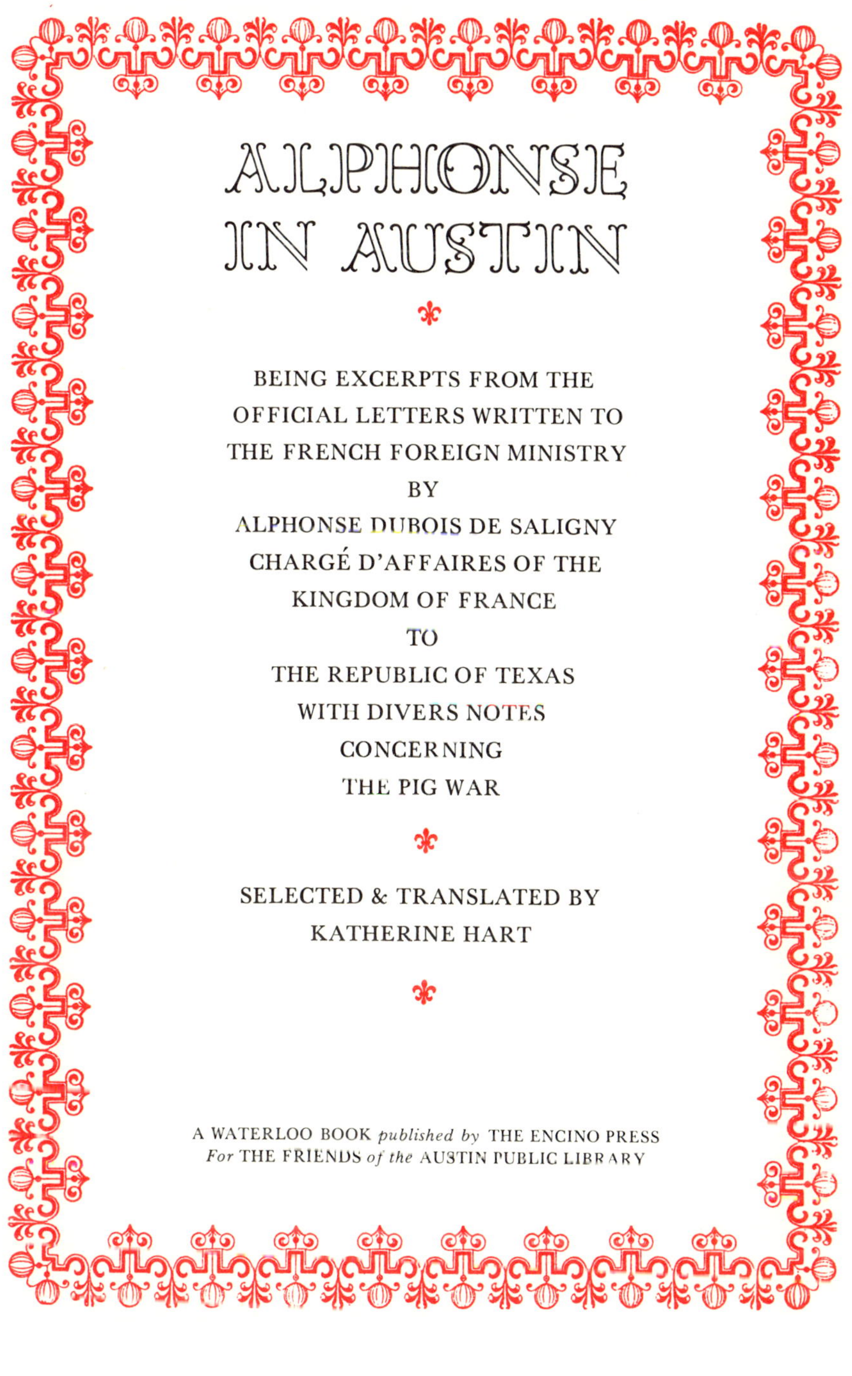

# ALPHONSE IN AUSTIN

BEING EXCERPTS FROM THE
OFFICIAL LETTERS WRITTEN TO
THE FRENCH FOREIGN MINISTRY
BY
ALPHONSE DUBOIS DE SALIGNY
CHARGÉ D'AFFAIRES OF THE
KINGDOM OF FRANCE
TO
THE REPUBLIC OF TEXAS
WITH DIVERS NOTES
CONCERNING
THE PIG WAR

SELECTED & TRANSLATED BY
KATHERINE HART

A WATERLOO BOOK *published by* THE ENCINO PRESS
*For* THE FRIENDS *of the* AUSTIN PUBLIC LIBRARY

WATERLOO BOOK

*Number 4*

Houston the 20 February 1839

M. the Count

M. Pontois having given me, in accord with the authorization made by Your Excellency in the dispatch of the 8 September last, the order to leave for *Texas* in order to fulfill here a mission of study and observation and to present to the Government of the King a report on the exact situation of this country, its chance of existence and its future as an independent state, its resources and the advantages that it can offer to our commerce and to our navigation, I set out from *New York* towards the end of November. I counted on being able to reach Houston before the end of the month of January; but I had not foreseen the numberless difficulties and the dangers of all sorts that awaited me on the road. In order to travel more quickly and more conveniently, I wanted to try to make part of the trip by water, but after having almost perished on two steamboats in the upper part of the *Ohio*, I had to renounce this way, very happy to have to regret only the loss of my baggage. Cold weather, whose equal cannot be remembered in the United States for fifty years, having from the beginning of December frozen all the rivers, made it necessary for me [to cross] the forests of *Indiana*, *Kentucky*, and *Mississippi*, in a country entirely without roads or means of regular transportation and almost completely uninhabited, traveling two hundred leagues in weather like Lapland, sometime on horseback, sometime in open cart, and obliged, most of the time, to spend the night in the middle of the woods. However painful the trip was, I cannot today regret it, since it has given me the opportunity to see with my own eyes a country up to now very little visited by foreigners, still very wild, but destined to play a great role and to exert an immense influence in the destinies of the American Nation, and since it has allowed me to make divers curious observations that I will have the honor of submitting later to Your Excellency. After a thousand hardships and a thousand dangers, the recital of which would not offer you the least interest, I arrived at the town of *Natchez* on the *Mississippi* toward the end of January. There I was forced to turn aside from my route to go to *New Orleans* to replace the baggage that I had lost and to procure horses and sundry necessary articles for my trip in *Texas*. The state of my health, which the fatigues and privations had greatly altered, forced me to stop some time in this city, and it is now only four days ago that I arrived in Houston, so overcome with weariness and so ill that I have been forced to go to bed and to stay in my room until this moment.

I had scarcely been at the seat of the Texian government two hours, when the President, who had been several months ago informed of my coming by General *Henderson* and who had the kindness to have reserved for me a lodging, an extremely difficult thing to procure in *Houston*, sent word to me that he looked forward with pleasure to seeing me and that he regretted knowing that I was sick. The same day, the members of the Cabinet and the Chief General came, individually, to write their names in my guest book. It was only yesterday that I was able to receive those among them with whom I had personal relations during my stay in the United States.

So recently arrived here and kept at home since my arrival, I would not be able to express yet any opinion on the country. What I can say as of today is that all the people that I have seen up to now, whether members of the administration or private individuals, agree in showing a true admiration and a great liking for France and that I, personally, have received from all sides testimonials of the most marked good will . . . .

I hope that I will in a day or two from now be able to go out and get to work, and I will have the honor of sending, next week, a first report on the result of my observations.

I cannot end this dispatch, M. le Comte, without thanking Your Excellency for the mark of confidence that Your Excellency has been willing to show me in charging me with a mission whose importance I appreciate. I pray you to be persuaded of the zeal with which I will fulfill the charge and of my efforts to render myself worthy of the goodness of the King and of the concern that Your Excellency has shown me.

I have the honor of being with respect,
M. the Count,
of Your Excellency
The very humble and very obedient servant
A. DuBois de Saligny.

Houston, 25 February 1839

Yesterday I paid a visit to M. the President. M. the Colonel *Bee*, secretary of State, had several times expressed the desire that his Excellency had to see me and had offered to introduce me. But to avoid having this visit seem in any way official, I preferred to be accompanied by

M. Labranche, chargé d'affaires of the United States. General Lamar gave me the most eager and cordial greeting.

At the end of a half-hour's conversation, during which the President seemed to take pleasure in finding the opportunity to show his profound admiration for France and for the King, I took leave of his Excellency who several times invited me to come to see him often.

As soon as we had left, M. Labranche could not keep from showing me his astonishment at the reception full of good will and warmth given me by General Lamar, whose glacial manners and the imperturbable silence in which he is accustomed to shut himself up from which nothing can wrest him, have caused him to be called the Dumb President.

Houston, 3 March 1839

M. Jefferson, when writing a letter in 1820 to one of his nephews who was leaving to continue his education at Cambridge University, outlined a course of studies and included these remarkable words: "One thing that I cannot recommend too highly to you is to give particular care to the study of Spanish; this language is the one that is spoken in an immense and rich portion of this continent, of which the Anglo-American race will be the mistress before a quarter of a century." Scarcely sixteen years have gone by, and the words of this great man of the American Nation, words which revealed a fixed idea of the government and of the people of the United States, are proved true, at least in part, by the events. Whatever opinion that is formed of the causes, obvious or hidden, which brought on the revolution of Texas, and whatever is thought of the more or less morality of the means which assured the success of the part that the United States, disdainful of the loss of neutrality, has taken indirectly or directly, there is a truth that cannot be denied; the Independence of Texas is henceforth an accomplished fact; this magnificent country is forever lost to Mexico.

For someone who has studied the Americans, who knows their restless nature, their insatiable avidity, their boundless ambition, their inclination for conceiving the most adventurous projects, their audacity, and their determination to carry out their projects, the establishment of the race on a part of the Territory of the Mexican Republic is a fact of the utmost gravity. In spite of the pretended moderation of the Texians today, and supposing that the peaceful protestations of their Government to Mexico are sincere, it would be a strange error if it were thought that, satisfied with the possession of the magnificent country of which they are today the masters, they would think only of finishing the conquest of the Indian, the bear and the buffalo, and that the energy that knows no obstacles, this admirable intelligence, this tireless activity which has already accomplished such miracles, should be used only for felling the forest and exploiting all their resources.

Houston, 8 March 1839

For some time the Indians, instigated by Mexico, have recommenced their incursions on the west of Texas. The news was received here last week that they had advanced in the number of 5 or 600 as far as the town of San Antonio de Bexar. Immediately the President requested the militia to move west and the next day several detachments forming a force of about 1200 men set out for San Antonio. This is more than needed to keep all the Indian tribes quiet.

The 2 of this month, the anniversary of the declaration of Independence of Texas was celebrated. The committee on arrangements wrote to invite me to be present at this ceremony. I answered the committee that I was personally grateful for the invitation but that I regretted that it was impossible for me to accept. Since, I have learned that General *Lamar*, himself, approved the circumspection with which I thought I should act in this instance.

I plan, Monsieur le Comte, to visit the towns on the seacoast; further, I intend to look over the principal settlements established on the Brazos, Colorado and Trinity rivers. I will then set out for Washington (D.C.), where I hope to arrive toward the middle of May and where I will have the honor of addressing Your Excellency a general report on the situation of this Country.

Houston, 16 March 1839

An immense advantage that the Texians have had is that, perfectly agreed among themselves on the means of accomplishing their revolution, they have also been agreed on the way to profit from it and to organize it. The Mexicans once chased from the Country, they set to work without delay; and without wasting time discussing theories, several days were sufficient for them to establish a government whose basis was already decided upon. Their Declaration of Independence was only a paraphrase, pale and without talent, of the famous manifesto drawn up, in its time, by Jefferson. Their constitution likewise was entirely imitative of that of the United States except for several modifications conceived in an even more democratic spirit . . . .

The election of General Houston was perhaps a just reward due the services that he had rendered to the cause of Texas in several important engagements and above all in the battle of San Jacinto, although, according to many people, this decisive victory should be attributed less to his military talents and to the orders given than to the bravery and élan of the troops, who even, if certain rumors can be believed, threw themselves on the Mexicans in spite of positive orders: but it [his election] was certainly a disastrous thing for Texas. General Houston was born with the happiest gifts, but, unfortunately, if nature did much for him, education did nothing at all. He is a tall and imposing man, his features are handsome and regular, he has a brave and resolute air; . . . his mind has much vivacity and penetration, his speech is easy and elegant enough although a little bombastic, and his whole person does not lack a certain dignity, in spite of the strange garb in which he is always muffled and the numerous rings of gold, silver and iron that he wears in his ears and on all his fingers. It was among the Indians, where he passed several years of his life, that he contracted this bizarre taste in dress: he also took their habits of laziness, of a wandering and vagabond life and, above all, their immoderate passion for strong liquors. While very young, he served with distinction under the command of General Jackson. Returned at the end of the war to his State, Tennessee, he succeeded, in spite of his disorderly life and debauchery, in getting himself elected a Representative to Congress, then Governor. During a trip that he made to Washington several years ago, he became, because of his misconduct and scenes of violence . . ., an object of scandal and contempt. He is one of the most discredited men in the United States. His presence at the head of the administration of this Country stopped a large number of rich and respectable families from coming to Texas, whose influence and capital would have given a great impetus to the Country. It also soon created a grievous embarrassment in the progress of the Government. Despite his deplorable reputation and his intemperate habits which he has not yet renounced and doubtless never will renounce, the prestige which surrounded him because of the victory of San Jacinto was such that the beginning of his administration was extremely popular; but soon his administrative incapacity, his excessive laziness, and his negligence toward his duties caused numerous complaints. The disdain he showed these complaints, the haughty and cavalier manner towards the two Chambers of Congress, and above all the stubbornness with which he persisted in his schemes (for Texas) to become a part of the United States, rejected by the majority of the people, raised violent opposition against him, and at the beginning of the last session, opened several days before the expiration of his Presidency, the misunderstanding between him and the Congress was so extensive that the operation of the Government was momentarily suspended. Today, now that he is no longer in power, he has become, as always happens, less unpopular; but in any event he has still few partisans . . . .

His successor General Mirabeau Bonaparte Lamar is an entirely different man. He does not possess any of those qualities which at first make an impression on the masses, awe them or win them over. He is little, ugly, awkward, and ordinary. Unbelievable effort must be made to pull from him the few words that he lets fall heavily and with difficulty. There is only one way to arouse him from his apparent lethargy, and that is to make some allusion to a war with Mexico: immediately his face lights up, his eyes blaze, his speech is short, quick, and positive and he starts to tell you with great warmth his campaign plans, his projects for conquest.

He is said to be a distinguished scholar and an elegant writer. Among his friends, who are very numerous, he is considered a capable man, enlightened, a general with a ready and firm eye, and a skilled administrator: what no one disputes, not even his most violent enemies, is the purity of his intentions, his integrity in every test and the sincerity of his patriotism. His courage and daring made him noticed in the battle of San Jacinto, and soon after his election to the Vice-Presidency, he distinguished himself by his opposition to the arbitrary and despotic measures of General Houston and principally by his opposition to annexation with the United States . . . .

. . . I have realized with some surprise, M. le Comte, and I take pleasure in telling Your Excellency that in this society, still so young, composed in large part of men of rough and almost fierce manners, of an ungovernable character and violent passions, where the Government does not have at its disposition either armed force or police, the people and property are better protected than in many of the States of the Union and the laws are more surely executed. This comes from, I think, the fact that the inhabitants of Texas, strongly influenced by the unjust prejudices against them that exist abroad, are making the most praiseworthy efforts to vindicate themselves in the eyes of other nations and are disposed, on every occasion, to assist justice. Those scenes of violence and murder so frequent in the States of the West and the Southwest are unknown here, and other crimes and offenses, comparatively very rare, are always promptly and severely punished.

One of the laws adopted by the last congress requires that, from the 1st of next October, the seat of government shall be transferred from Houston to the West on the banks of the Colorado, a place even today ravaged by the Indians. The Commissioners who left to choose the location where the new capital is to be built have not yet made their choice, or at least the government had not yet received their report. General Lamar expects the law to be executed on the appointed day, and he has decided to set out for the West at the end of September, carrying his tent with him, and followed by his cabinet and employees from various departments. Indeed, this is what already took place when the government was established at Houston about 20 months ago; and on this subject I shall show Your Excellency what kinds of men inhabit this country. At that time there was not a single house in Houston. On the banks of the Bayou (little river) of the Buffalo, trees were chopped down from which was hastily constructed a log house or poor cabin composed of only one room, where the President and the members of the cabinet were lodged. Then, beside it was built a sort of vast barn also made from the trunks of trees, roughly piled up, where Senators, Representatives and public officials were crammed in, pell mell, only too happy to find there a shelter against the winter and to be able to procure there, by paying in gold, not beds, but sacks filled with shavings into which they put themselves, as though into furs, to pass the night. What a race! What can the Mexicans do against men of this character!

The latest news from the West is of a satisfactory nature. The Indians, beaten in several encounters, have disappeared. The establishment of the seat of Government will have the inevitable effect of pushing them much further west.

Houston, 26 March 1839

For ten days the *Texians* have given themselves up to transports of the greatest happiness: banquets, patriotic meetings, military reviews, and public rejoicings of all sorts have followed each other without interruptions. All this joy is caused by the presence of General Hamilton, who arrived here last week accompanied by several important citizens of South

Carolina, Mississippi and Florida. The character of General Hamilton, the fine reputation that he enjoys in the United States, the influence that he has in the South, and the zeal with which he embraced the cause of Texas from the beginning and has sustained ever since, give this visit great importance; thus the *Texians* have received him with extraordinary honors.

. . . for a year the new Republic has made unbelievable progress: its resources, immense in the future, begin to be better organized; its credit is being established; and its population which is intelligent, active, hardworking, and resolute receives numerous reinforcements each day; and the number of immigrants coming just by way of Nacogdoches, during the last months, has reached 7,535.

Velasco, 17 April 1839

As I told Your Excellency in my last dispatch, I took advantage of the President's offer to accompany him in his excursion on board the Zavala. I visited in turn the ports of Galveston, Velasco, Matagorda and other coastal towns. Familiar as I have been for some time with the sight of the marvels brought forth each day by the activity of the Americans, and although I am accustomed to seeing on every side the flourishing towns spring from the earth, as if by magic, in places where even yesterday rose thick forests which were impenetrable to man and occupied only by wild animals, I swear, Monsieur le Comte, that I have never seen anything comparable to the sights which were offered to me on the trip that I have just made. I did not yet know what prodigious feats could be accomplished by the audacious industry of this American race, by their tireless perseverance, and by their energetic spirit of enterprise. This magnificent land, owned for several centuries by the Spanish race, was even two years ago only a vast solitude where the traveler scarcely found, here and there, a few poor villages, some miserable huts of fishermen. Today, one finds towns everywhere, all located with a rare widsom and, as in most American towns, built on a large plan. Everywhere there is digging, construction, improvement of ports and roads, and deepening of river channels. It is impossible not to recognize right away a country on a rapid path of progress and destined to an immense future. Matagorda and Velasco, both situated on the Gulf of Mexico, the first on the bay of the same name and the second at the mouth of the Brazos, are already two important ports that have a fairly extensive trade with the United States. But the principal port of Texas is Galveston which is situated on the island and bay of the same name. It is there above all that the enterprising genius of the Americans shows that it will not draw back before any obstacle. The island of Galveston is nothing but a sand bank, 30 miles long, from one to five miles wide, and so flat that at a distance it resembles an enormous raft floating on the sea. The sterility of this island, on which there is only a trace of vegetation, and the risk that it runs of being submerged at any time by the winds from the north and the south have always repulsed those who have tried to come to stay there. Lafitte, the famous French pirate, was the first who dared to locate there. The very dangers which surround it persuaded him to choose it for his headquarters. In 1819, the French from the Champ d'Asile, repulsed by the Indians and the Spanish, took refuge there, but a terrible storm, which lasted three days, destroyed all the settlement that they had built and obliged them to embark hurriedly for the United States. Only a spirit of speculation could have dreamt of choosing this island to make of it the entrepôt of Texas. Several individuals combined to buy a part of the island at a low price and have built the foundations of a town which, started at the most 18 months or two years ago, already numbers 7 to 800 houses. It has undergone three floods, but this has not discouraged the hardy inhabitants who only pursue with more ardour their work of improvement and who are getting ready to build an immense levee to protect themselves from the sea. Nevertheless, despite their intrepidity and perseverance and in spite of the help of some of the richest capitalists in the United States, (Mr. Nicholas Biddle, among others, is personally interested, it is said, to the extent of several hundred thousand francs in the town of Galveston), I do not know whether they will succeed in their magnificent plans. It is doubtful whether foreign commerce will consent to expose its merchandise to the dangers that constantly threaten this town. I would not be surprised to see other speculators set about creating a rival town on some point of stable land on Galveston bay, the only bay in this Country, it seems, which large merchant ships could enter.

The commerce of Galveston with the United States is already so considerable that five steamboats are exclusively occupied in maintaining the service between this port and New Orleans. Another steamboat line to Mobile has just been organized. The town of Houston, as Your Excellency knows, is joined to Galveston by the small Buffalo river, which empties into the northwest end of the bay. Daily six steamboats of 250 to 300 tons go up and down this river, which in many places is no wider than the Bierre river, but which has an average depth of 20 feet. It is a curious sight to see them boldly, but not easily, make a passage through the branches of the sycamores and the magnolias that interlace from one side to the other of the Buffalo, to see them graze at every moment the two banks, and to see them continue their way leaving the river covered with broken branches of trees and their own debris. The establishment of another steamboat service on the Sabine has been announced. It is said that one steamboat has gone up this river a distance of 400 miles.

The Texians are not limiting themselves to exploiting the means of communication that nature has put at their disposal. They are preparing to create others destined to make life circulate in the richest parts of their Country. Two railroads, their object to unite the Colorado and Brazos rivers with the city of Houston and Galveston bay, have been planned. They will soon be started, and it is hoped that they can be finished 18 months from now.

At Brazoria, a pretty little town on the Brazos river, where I spent two days at the home of one of the wealthiest planters in the Country, several people offered to accompany me if I wished to make an excursion to the west, and I accepted eagerly. Ten of us left on horseback, well armed for fear of Indians, for San Antonio de Bexar, capital of the province under the Spanish rule. A detailed description of my trip and of the magnificent country that I went through would have no interest for Your Excellency and besides would be out of place in this dispatch. Thus I will content myself, M. le Comte, by telling you that I found the country watered by the Brazos and the Colorado superior to all that I have heard about it. There are on these two rivers a considerable number of fine plantations which are admirably developed and would not be surpassed in any way by the majority of those in Mississippi and Alabama. The chances for the next crop seem very favorable. It is said that they will produce about a hundred thousand bales of cotton, which is enormous if one realizes that it is scarcely two years since the Texas Revolution ended. The success, obtained by the Planters who were the first to profit from the privilege granted only to Americans to bring their slaves into the Country, has been so complete that those (Planters) in the two Carolinas and Georgias, who are tired of cultivating worn out land which only gives small revenue, are leaving them to come settle on virgin soil where they hope to find bountiful harvests. Probably this example will be followed by a great number of planters from Maryland and Virginia. Finally, in Mississippi which a financial crisis has thrown into a complete state of revolution, some inhabitants, who are burdened with debt and pursued by creditors that they cannot satisfy, have opposed (with guns in hand) the seizing of their slaves and have brought them to Texas, leaving as the only guarantee to their creditors the land which had not been paid for. While thinking about the different reasons which very soon must bring a large part of the population of the Southern States to this country, I understand that there is no exaggeration in the opinion that I have frequently heard advanced even by planters from the United States, that ten years from now the cotton produced by Texas will equal half of the crop in the United States.

Several planters of the Brazos River have tried the cultivation of the mulberry tree with much success. I believe this cultivation and the breeding of the silk worm is destined to be greatly developed in this country.

After ten days of interesting but difficult travel, we arrived at our destination without any noteworthy incident except the meeting, near San Antonio, of several bands of Indians who, although ten times more numerous than we, fled at our approach. San-Antonio-de-Bexar is an old city, built by the Spaniards and having about 1500 inhabitants. It has lost a great deal of its importance in the last years, but it is beginning to revive somewhat since it has become the entrepôt for a considerable trade with the inhabitants of New Mexico, who show a very friendly attitude towards the Texians and who bring trains of horses and beasts every day to San Antonio to buy goods manufactured in Europe and the United States . . . .

My attention, M. le Comte, has also been directed to another point. One finds on several rivers of Texas, particularly on the Brazos, vast forests of wood suitable for building. The *live oaks* (green oaks) are very abundant here, which the Americans have used almost exclusively for their navy up to now and whose superiority over every other wood has been recognized the world over . . . .

I stayed in San Antonio only long enough to rest a little and returned here by almost the same route used going there. I arrived at this town yesterday and intend to go to Houston day after tomorrow.

In addition, M. le Comte, I have found their country enjoying the most perfect tranquility and order. If there had remained in my mind any doubts on the situation of the New Republic, on its resources and the means by which it could maintain its independence, this trip that I have just made would have dissipated them. Everywhere I saw people a little uncivilized, perhaps, and sometimes crude, but intelligent, industrious, and determined, who, at the first signal, would all run to arms and against whom all the forces of Mexico would not be able, I believe, to hold out for so long.

I met in all my trip with the greatest hospitality and the most extraordinary consideration which was directed much less toward me, without any doubt, than to the government of the nation to which I have the honor to belong. At Velasco, Matagorda and Galveston, the citizens wanted to give me public dinners. But, though I expressed to them a deep gratitude, I excused myself and made them understand that my position required extreme reserve, which even their interest demanded that I exercise, for fear that if the report that I have to make on their Country to the Government of the King should be favorable, one might in some way see in it the reflection of the good treatment of which I should have been the object.

Houston, 20 April 1839

These people, M. le Comte, are unbelievably inexperienced and ignorant in all that concerns external politics. They have no idea of the respective situation of different nations, of the rules and the usages that govern the relations amongst these nations; besides, being Americans, they have boundless vanity and presumption . . . .

Houston, 1 May 1839

I intend, M. le Comte, to set out for Washington in four or five days. From there I will have the honor of making you a last report of my observations on Texas.

New York, 24 June 1839

I have the honor to announce to Your Excellency that I left Texas the 5th of last month and that I have been back in this city for several days.

The mission with which I was charged, M. le Maréchal, had for its aim to gather exact information on the situation of Texas, its chances of existence as an independent state and the outlets it could offer for our commerce and our industry. I endeavored, in the dispatches that I had the honor to address successively to M. le Comte Molé, to furnish to the government of the King the most precise information on these different points . . . .

I have already said, and I repeat it here, M. le Maréchal, Texas is forever lost to Mexico.

If I pass now to the internal situation, I see that it is not less satisfying, in spite of certain difficulties inevitable among a people in its cradle. Order has been established in this country with admirable rapidity. The authority of the law is everywhere recognized and respected: foreigners can come there without fearing any of those insults or outrages, those vexations of all sorts to which they are so often exposed in Mexico and in the greater part of the Republic of South America, and the powers that establish political relations with the new state will find in its government the will as well as the power not to suffer in its territory any infringment of the fixed laws of justice and of the principles of the rights of nations.

The details contained in my preceding dispatches have been able to give Your Excellency an idea of the prosperity which this newly born republic enjoys and the extension each day of its commerce.

I repeat in closing, M. le Maréchal, that the recognition of the Independence of Texas by the Government of the King can assure great advantages to France for a long time. It is a fine opportunity that is offered to us; we cannot let it escape.

Greensborough (Georgia) 3 January 1840

M. le Maréchal,

Having left England the 16th of last November aboard the steamship *Liverpool*, I arrived at New York the 5th December. General *Henderson*, who had preceded me by only a week, when departing to visit his family in North Carolina, had left me a letter announcing his return soon to New York and asking me to wait for him there in order to continue our journey to Texas together. But several days later a second letter from him told me that the premature rigor of the season and the fear of not being able, if he waited any longer, to descend the Ohio river and the Mississippi had determined him to continue his voyage directly without returning to New York. Immediately I set out myself, hoping that by exercising great dispatch I would be able to rejoin General Henderson either at Cincinnati or at Louisville. But once again I had to learn, at my expense, that in the United States, this country famous for railroads, canals, and steamboats, one does not always travel with that extreme ease, with that marvelous rapidity, that they boast about so much in Europe.

The Americans possess, it is true, from one end to the other of their immense country, means of transportation more numerous, and more rapid than any other country in the world. But no general idea governs the exploitation of these magnificent rivers, these numerous railroads, and these endless canals that cross it in every direction. The foresight of each company stops at the boundaries of the territory worked by it, without its deigning to bother the least in the world about what will happen to the travellers beyond, or even to take the slightest precaution to prepare for inevitable accidents. It also happens very frequently, especially in winter, when the railroads find themselves suddenly loaded down with snow, the rivers and canals closed by ice, that the unhappy travelers find themselves for weeks, even whole months, bottled up in the middle of a deserted country, not able to go forwards or backwards. When, in the month of December 1838, the steamboat on which I had taken passage was completely trapped in the middle of the ice of the Ohio river, the captain came, with an admirable sang-froid, to warn me and about a hundred other passengers that we would probably be obliged to remain on board until the month of March and that he would neglect nothing to treat us as best he could during all this time and to make us happy! One understands, up to a certain point, that in the new states like those in the West the companies do not have at their disposition all the necessary resources to assure service in case of accident; and one must even admit that it is already a great accomplishment for them to have succeeded in so little time in endowing a country which, only yesterday, was just a vast wilderness, with these long lines of railroads and with steamboats. But the same is found in the oldest and most advanced states in the Union; the same neglect of the most simple precautions, and the same complete indifference to the well-being and interests of the travelers are noticeable to such an extent that it is hard to understand, in a nation so universally tormented with the need for locomotion and to whom time is such a precious capital; and to me what is even more extraordinary is the patience, the resignation of the public, who submit to the most troublesome caprices, the most unreasonable demands without offering the least complaint and without allowing themselves the slightest murmur.

As for me who, during my stay in the United States, has had many occasions to curse the carelessness of companies, captains and agents for steamboats, railroads, etc., etc., never have I suffered as much as since the beginning of this unlucky voyage. An accident happening on the railroad from New York to Washington caused me a delay of four days in a journey that is made ordinarily in eighteen hours. Then, on my arrival in Louisville, after a thousand anxieties and a thousand hardships, I found that . . . the steamboats could no longer descend the river. It was then necessary for me to decide to take another route and, first of all, to retrace my steps a distance of about 400 miles. According to the various information that I obtained, I decided to go to Mobile by way of Virginia, the two Carolinas, Georgia and the route recently opened across Florida. I had a great deal of trouble getting to Wilmington in North Carolina, where I took a steamboat, to go to Charleston, which was almost lost in the midst of a storm off Cape *Fear*. When I arrived at Charleston, news had just been received that the Indians in Florida had, several leagues from *Tallahassee*, attacked seven public coaches and massacred the travelers that they carried. This news decided the company of the route through Florida to suspend its operation immediately; and no other course remained for me to take except to reach the Alabama river by the roads of the interior of Georgia which the bad weather had made impassable. Thus a week after I left Charleston, I am scarcely yet a hundred leagues from it, obliged, I fear, in order to procure means of transportation, to remain several days in this village, where I arrived yesterday, not knowing how to continue my trip.

These continual delays, M. le Marèchal, vex me more than I am able to express to Your Excellency. I consider it essential that I arrive in Austin before the adjournment of Congress, and if I must encounter at the end of my trip the same obstacles as up to now, I fear I will not succeed . . . .

During my passage through Augusta, I scanned several fairly recent newspapers from Texas. It seems that yellow fever, which up to now had spared that country, made frightful ravages there last summer and autumn. Galveston and Houston have especially suffered from this scourge which did not spread into the interior beyond this latter town. Although the doctors maintain that the disease did not originate in Texas and that it was imported there by immigrants come from New Orleans and Florida, a great number of the inhabitants of the coast have hurriedly left the country and the immigration movement from the United States has for the moment slowed down. But now that the scourge has disappeared and the fears have been dissipated, this movement has recommenced, and the immigration proceeds with more enthusiasm than ever towards the fertile regions of the young Republic.

Houston, 19 January 1840

Here I am finally in Texas. I arrived day before yesterday at Galveston and only stopped there a few hours to wait for the steamboat that brought me here. General Henderson, who arrived at Galveston five or six days before me, hurried to dispatch a courier to Austin carrying the Treaty signed by Your Excellency and him, the 25th of last September and he, himself, set out two days ago for the seat of government.

The recognition of the new Republic by the government of the King has caused universal joy in this country. I was awaited impatiently and have been received with the greatest testimony of respect and gratitude toward France and of goodwill for me personally. At Galveston and at Houston artillery salvos saluted my arrival and the authorities and the principal inhabitants came to compliment me and to offer me, in the name of their citizens, a great public dinner. I expressed to them my sincere appreciation and excused myself from accepting their invitation because of the obligation that I had to go to Austin without delay. It appears that the session must last until the early part of February, and, in spite of the bad condition of the roads, I hope to arrive at that town before the adjournment of Congress. If I cannot procure horses today, I will set out anyway tomorrow.

The Comanches, the most belligerent and powerful of the Indian tribes in Texas, have just made friendly overtures to General Lamar and asked him to send to San Antonio de Bexar, towards the middle of next month, Commissioners for the purpose of talking with their chiefs on the basis for an agreement. The President hastened to accept these proposals in which many persons here see only a trap on the part of the Comanches.

Scarcely had the answer of General Lamar to this Indian tribe become known when the Lipans and the Tonkaways, mortal enemies of the Comanches and for three years allied with the Texians to whom their assistance had often been very useful, made it known that they did not wish to remain friends any longer with a government disposed to treat with their own enemies and that since Texas wished to make peace with the Comanches, it must expect nothing from the Lipans and the Tonkaways except a war to the death. Joining the deed to the message, they immediately spread out in the country and commenced their depredations.

P.S. 20 January—

Today travelers bring news from Austin, which they left the 15th. In its meeting of the day before, the Senate approved, unanimously by the members present, the Treaty of the 25th September.

It seems that the roads from here to Austin are infested by Lipans and Tonkaways, who, scattered through the country in little bands, ravage the isolated inhabitants, intercept their communications, stop and massacre travelers. After this news, people tried to dissuade me from leaving, but it is absolutely necessary that I get to Austin before the recess of Congress, and I have decided to set out tomorrow. I will be forced only to take an escort of six or eight men.

Austin, January 30, 1840

As I had the honor of announcing to Your Excellency by my last dispatch, I left Houston the 21 of this month. I did well to have myself accompanied by an escort, for the third day of my trip I fell into the middle of a fairly considerable party of Indians who at first appeared to want to attack us or, at least, to take from us some horses and a part of my baggage, but seeing my small troop, well armed and ready to offer vigorous resistance, they suddenly withdrew. They probably took us for some detachment of troops sent to punish them for, although ten times more numerous than we, they rapidly crossed back over the Colorado and disappeared into the mountains.

After five days of hard travel on roads that the rain had made almost impassible, I arrived at this residence, two or three days ago, without other accident save the loss of two of my horses, one of whom died of fatigue in the middle of the trip and the other drowned in crossing the Brazos river.

The welcome that awaited me here, M. le Maréchal, was not less cordial or flattering than that I received at Galveston and Houston. At the news of my approach, several members of the Cabinet, the city officials and a crowd of army officers and employees of different Departments hurried to come on horseback to meet me. I met, five or six miles from the city in the magnificent valley of the Colorado, this procession having at its head General Henderson and the Mayor. The latter addressed me, in the name of his fellow citizens, a discourse full of the greatest praise for France and its Government. I answered as best I could and then went, escorted by this cavalcade, to the lodging that they had been kind enough to engage for me.

In the course of the day, I received visits from the Vice-President, all the members of the Cabinet and the principal public officials.

The welcome that I received in this country, M. le Maréchal, therefore leaves nothing to be desired. Everybody demonstrates a sympathy and preference for France, the more decided because the spirits are all very irritated against England, and everything makes it hopeful that each day will tend to fortify the newly established bonds of friendship between the two countries.

Austin, 15 February 1840

There is still no mail service established between here and Houston, and one is obliged most of the time to make use, for the sending of dispatches, of special opportunities very rare and undependable, especially now when the country is infested with Indians who, spread through the country in little bands, stop and massacre travelers. Several days ago they advanced to within two or three leagues of Austin, pillaged a poor isolated cabin, killed the owner who only had an old Negro with him to defend himself, and led away some thirty horses and mules. It would not have been very prudent to commit important documents to such hazardous risks, and I decided to send the Treaty by a trustworthy man, intelligent and resolute, that I brought from France with me. I will have him accompanied by two guides, one an Indian and the other Mexican, both of them devoted for a long time to the cause of Texas, and I hope, thanks to this precaution, that it will arrive without accident at Galveston. I commanded the messenger to go as far as New Orleans in order to avoid having the Treaty mislaid or even lost. One cannot imagine the negligence of the captains of the steamboats and the postal agents at New Orleans: I have not been able to discover any trace of two parcels containing important letters which were sent from that city to my address here last December.

They are not taking any precautions to shelter their new capital from attacks, which with no means of defense save the courage of its inhabitants, whose number has been reduced by a third since the adjournment of Congress, could not only not hold out long against a Mexican army, but would have a lot of trouble resisting four or five hundred Indians if they took the idea of striking it unexpectedly.

It is true that since the uprising of the Lipans and the Tonkaways, who though very troublesome are not numerous enough to be dangerous, the Texians are much more tranquil concerning the Indians. The Comanches, the most fearsome of the Indian tribes of Texas, due to their consistent policy and their implacable hatred of the Lipans, have shown an extreme good will to the Texians since the recent depredations of the Lipans. Ten of their chiefs have already arrived at San Antonio where a meeting is to take place around the middle of March, . . . between the principal chiefs of their tribe and the commissioners sent by General Lamar. What will result from these conferences and how much should the Texian Government believe in the sincerity of these savages, whose good faith is not their outstanding trait and of whom one should never be more suspicious than when they are prodigal with their protestations of friendship? This is what events alone will prove.

Austin, 10 March 1840

The Lipans last week killed several travelers near Bastrop on the Colorado River, forty miles from Austin. Several detachments of volunteers have just been sent to pursue them.

The French ship the *Only Son*, sent from Marseilles by the Fitch Co., arrived at Galveston the 25th of last month. It is the first ship to come directly from France to Texas. Let us hope that others will not delay in following it.

Judge Lipscomb soon intends to make an excursion to Matagorda and to Velasco. I intend to take advantage of the offer that he made me of accompanying him. Perhaps I shall decide to go as far as New Orleans where I want to oversee the sending of my baggage myself which has just arrived there. Besides I need to make arragnements with the post office of that city and the steamship companies to insure the conveyance of my correspondence which has been conducted in the most irregular way up to now.

Houston, 17 March 1840

I left Austin the 12th of this month and arrived at Houston this morning. The vigorous measures adopted by the Government against the Indians have had a salutary effect. Communication has become easier and more dependable. I did not have any unpleasant encounter on the road; it is true that I traveled as far as San Felipe de Austin, a town situated on the Brazos River 60 miles from here, with a detachment of 50 Texian volunteers, which was sufficient to keep the Lipans and the Tonkaways at a respectful distance.

I succeeded en route in obtaining news of my messenger carrying the Treaty. It seems that the bad condition of the roads, and above all the presence of Indians forced him to make a long detour and to go by Matagorda. I hope that he will have been able to get there without accident.

Two months have scarcely passed. M. le Maréchal, since I traversed, on my way to Austin, the country that I have just passed through again, and I was astonished by the improvements that have taken place in this short time. Fine farms and pretty villages have risen in places which were still only a vast solitude at my last passage.

During the last session, the opposition has sharply attacked the choice of Austin for the location of the Capital of the Republic. It has bitterly reproached General Lamar, one of the prime movers of this measure, for having transferred the seat of Government to the edge of the country in a land inhabited only by Indians, exposed to their continual incursions and even to the attacks of

Mexicans, far away from the heart of the population, deprived of all communication with the centers of commerce and with the sea, and where one can obtain only at the price of gold the barest necessities. One would not be able to deny that there is some truth in these accusations, but although the Government can, perhaps with some reason, be blamed for acting a little prematurely, I think nevertheless that the policy which made it settle at Austin was clever and far-sighted, as well as resolute and bold. To me, it was the only way to protect from the incursions of the Indians the richest part of the Country where, up to now, immigrants hardly dared to venture, and which, under the ax and the plow of the industrious and intelligent population who are coming there in crowds, will soon become an inexhaustible source of riches for Texas. Last January, I camped to spend the night in a pretty valley not far from Bastrop, where there had never been a plow. On my return, I found there superb fields of wheat and corn: more than 800 acres of land have been put in cultivation there since that time. It would be impossible to have an idea of the rapidity with which the upper parts of the Colorado and Brazos valleys are being settled. It is there that come today the long waves of immigrants whom the United States send forth towards Texas each day. A large part even of those who settled on the Sabine and the Trinity are leaving their less rich regions to go to the West. To tell the truth, this is the secret of the hatred of Austin by the opposition. Everything is a matter of speculation for the Americans. Politics, morality, religion even—they know how to exploit them all, to make a deal out of them. Those in Europe who ceaselessly praise the excellence of the *American system*, who each day represent this Government as the only one into which fraud and corruption cannot be introduced, and who cannot find strong enough terms to celebrate the purity, the morality, the disinterestedness, and the patriotism which these *model institutions* germinate in all classes of society, would be sorely surprised, even cruelly disappointed, if they were initiated into all the means of corruption put into practice by the parties; into all the evil influences which agitate those political assemblies that one would like to represent to us as the personification of all the noble and elevated passions; into those shameful transactions and those underhanded dealings which are consumated each day in the dark. It is pretended that some of the people, who are attached to the Government and who possess its confidence, let themselves be guided solely in the choice that they made of Austin by considerations of private interest; they are accused of having been engaged in lucratic speculations that were not too honorable. These accusations, which are not made of General Lamar whose integrity and disinterestedness is recognized even by his enemies, can be justified in respect to certain individuals. . . . The most influential leaders are all big landowners on the Sabine, the Trinity, and the lesser streams in the eastern part, each having in his pocket his own plan for a capital, and one must expect to see them oppose a measure which ruins all their dreams of fortune and must have the inevitable effect of considerably diminishing the value that they attach to their property. Thus they did not neglect anything in the last session that might change again the seat of Government. But all their efforts have been useless. Doubtless they will renew their efforts in the next Congress, without more success probably, not because it would be impossible to get together a majority determined to leave Austin but because divided as they are by personal interest, they will not succeed, I believe, in agreeing on the choice of another capital.

The *Cherokee* Indians, chased from their Country by the United States, sent a deputation to the Government of Coahuila to ask it to give to their tribe a grant of land in this State. The Government has refused to admit the whole tribe, but it seemed disposed to allow 200 families to come settle on the territory situated between San Fernando and Santa Rosa on the right bank of the Rio Grande. The Indians, of whom the Federalists wished to make, as it were, a sort of advance post against the Government of Mexico, did not wish to accept this proposition.

I intend to leave in a day or two for New Orleans. I shall travel with General Samuel Houston, who is going to be married in the United States, although he is already married to a woman whom he abandoned fifteen years ago in Tennessee. It is true that, not having been successful in obtaining a decree of divorce in the United States, he took the step of having his marriage annulled in Texas by an act of Congress, the legality of which the American authorities could very well, it seems to me, refuse to recognize. It would be rather curious to see the Ex-president of Texas prosecuted in an American court for the crime of bigamy.

New Orleans, 30 March 1840

A bloody tragedy has just taken place at San Antonio de Bexar. Letters that I received from this city on my arrival here several days ago give me particular details of this unhappy affair which I hasten to pass on to you.

Your Excellency knows that General Lamar, giving ear to the peaceful overtures of the Comanches, consented to an interview's taking place, towards the middle of March at San Antonio, between the Texian Commissioners and the principal chiefs of these Indians for the purpose of examining the propositions that the latter wished to make to Texas. Nevertheless the cabinet at Austin made one preliminary and *sine qua non* condition to these talks. The Comanches must, before everything, bring to San Antonio thirteen Whites made prisoners by them about eighteen months ago and whose ransom must be fixed by private contract; and they were notified that unless this condition were fulfilled, no proposition of theirs would be heard and that consequently it would be useless for them to come to San Antonio.

The 19th of this month, twelve Comanche Chiefs, escorted by some fifty Warriors and a certain number of women and children, arrived in that town. Contrary to the word that they had given and in spite of the formal declarations of the Texian Government, they only brought with them a young girl, Miss Lockhart. They pretended that the other prisoners were dispersed in the mountains far away, but that if they were paid a generous ransom for Miss Lockhart, they would have them all come, but one by one in order to bargain for the ransom of each one separately.

Asked about the fate of her companions in misfortune, the young girl declared that it was false that they were far away as the Indians pretended and dispersed in the mountains, and that she had seen them all at the Indians' camp the night before her departure. Colonel W. C. Cooke, commandant at San Antonio, indignant at the duplicity of the savage chiefs, resolved to secure the Indians and keep them as hostages for the prisoners. Consequently, he had two companies advance under the command of Colonel Fisher and placed them near the house where the conferences were to take place. Then having come into the hall where the Indian Chiefs were, he reproached them severly for their lack of faith, and after a fairly violent discussion, having had a company enter into the assembly hall, he declared to the Comanches that they were his prisoners and would not be given liberty until the Whites detained by them should have arrived safe and sound at San Antonio. At these words, one of the Comanche Chiefs hurled himself towards the door to leave, but the sentinel who was guarding it having crossed bayonet, the Indian drew his dagger and struck him with it. At the same time, the other Chiefs also hurried towards the door and wounded several Texians. A terrible fight started, and the twelve Indians were massacred in a few minutes.

At the noise of the fight, the company remaining in the courtyard under the orders of Captain Reed was attacked with fury by the other Comanche warriors who, pushed back into the neighboring houses, barricaded themselves there and rained a hail of bullets and arrows on their enemies. It seems that these savages fought with a fury and despair that cannot be imagined, and everywhere that their arrows carried, they penetrated up to the feathers with which the wood is furnished. Their women above all, it is said, made themselves remarkable for their fury. A warrior, having fled alone into a house, refused with indignation the offer that was made him to leave him his life if he consented to surrender. The assailants, not being able to enter the house where he was entrenched and from which he had killed and wounded several of them, decided to set fire to the house, and when this bold warrior tried to leave to escape the flames, he fell pierced by bullets. But in spite of the courage shown by the Indians, the struggle was too unequal to be prolonged. Their most redoubtable warriors having succumbed, those who remained were made prisoners. Some of them succeeded in making their retreat and in getting to the other side of the San-Antonio river, but, pursued by Colonel Wells' cavalry, they were almost all massacred. The number of dead on the Comanche side is estimated about fifty, without counting several women and children. A dozen prisoners besides remained in the hands of the Texians. They had 7 dead and 8 wounded. Among the first named was Judge Thompson, a young man of merit who was to have been sent as Secretary to the Legation to Paris. At my passage through Houston in the month of January, I succeeded in persuading him to accompany me to Austin, and once there he could not resist the desire to go to San-Antonio to be present at the meeting of the Comanche Chiefs.

This bloody catastrophe produced, as Your Excellency can imagine, a profound sensation in Texas and the United States. But the role played by the Texians in this deplorable affair was not judged there as severely as some people may in Europe. The population that lives in the vicinity of the Indians has had to suffer much from their perfidy and their treachery. These savages have so often taken advantage of the truces that have been granted them to come with peaceful protestations and appearances of friendship, only to plunder and murder the poor wretches who, relying on the faith of the treaties, received them as friends, that there exists toward them a violent hatred and exasperation, which often knows no bounds and does not always show itself too scrupulous on the more or less degree of the morality or the means employed to fight them.

As soon as the Tonkaways and the Lipans were told of what happened, they came to put themselves at the disposition of the Texian officers to march against their mortal enemies, the Comanches. Now the Governmént is easy on the side of these two tribes.

New Orleans, 27 April 1840

The Comanches persist in their ideas of war against Texas. Several bands of them have shown themselves about twenty miles above Austin and have had divers engagements with weak detachments of volunteers. The advantage has always remained with the volunteers although they were ten times less numerous than their enemies. The Indians will not be able to hold out before the forces that arrive each day in the West, and they are doubtless going to feel the necessity of withdrawing into their mountains, not to reappear until the country shall be less well protected and when they think they will be able to take up the pursuit of their depredations without danger.

Although my presence is in no way necessary in Austin, I was getting ready to return there in a few days. But I learn that all the members of the government have left this place of abode which is deserted today. Judge Lipscomb, who continues since he was named Secretary of State to practice his profession of lawyer, is busy pleading a case in Washington. The Secretary of War is on his Brazos river plantation, and the Attorney General, Judge Webb, has left for his land on the Colorado. Finally, the President is expected daily in Galveston where he is to pass some time. I have thought under these circumstances that I could put off my departure for a while in order to build up my health a little, which has been considerably impaired. I am also not displeased to wait for General Samuel Houston, who with his young wife is going to visit the country watered by the Trinity and the Sabine and who has invited me to accompany him. This trip will be an opportunity for me to study, and I will have the honor of reporting to Your Excellency the observations that I will be able to make.

New Orleans, 4 May 1840

I set out tomorrow or the day following for Austin by way of Natchitoches and Nacogdoches. I hope, as I have told Your Excellency, to make this trip with the former President, General Samuel Houston.

New Orleans, 17 May 1840

I left New Orleans the 6th of this month to return to Texas by the Red River and the interior. But arriving at Natchitoches, the state of my health obliged me to stop and not having the strength to continue my route by land where I would have to travel horseback, I had to decide to return here. I have engaged my passage on a steamboat leaving day after tomorrow for Galveston.

Moreover, I do not need to regret my excursion to Natchitoches, for it enabled me to make some useful observations and to recognize the truth of the reports that are given me everyday on the prodigious movement of immigration coming to Texas. From the point at which the Red River flows into the Mississippi, I found the road thronged with carriages, wagons, horses, mules, and beasts. They were residents of Mississippi who were moving to Texas with their wives, their children, their Negroes, and all their furniture. The convoy, composed of 4 to 500 wagons, counted more than 1800 persons of whom about 300 were Negroes and covered a space of nearly three miles. One could have called it an entire village immigrating to a better climate. It seems that, since the end of winter, the roads are covered with similar convoys. During the three days that I stayed at Natchitoches, I saw more than 300 wagons pass loaded with men, women and children. Trustworthy persons have assured me that the number of immigrants who have passed through this town to go to Texas during the month of April only is no less than 8,000!

The newspapers announced the arrival of General Lamar at Galveston, who must remain there about a month to build up his health—

Galveston, 6 June 1840

I arrived here seven or eight days ago, and yesterday I went to pay my respects to the President who received me with his usual kindness and showed me, in the most flattering terms, the pleasure that he felt in seeing me again. The health of General Lamar seems gravely altered, and although his condition is in no way yet alarming, his friends fear that it will take much time, care and rest to be completely restored. He intends to pass the greater part of the summer at Galveston in order to take the sea baths there that his doctors strongly advise.

The new chargé d'affaires of the United States, M. Flood, just arrived in this country and has already found the means to make himself very unpopular by his disorderly and intemperate conduct which makes a striking contrast with the deportment, a little stiff and stern but always dignified and correct, of his predecessor, M. Labranche. The Texians are furious that the United States has sent them such a minister who dishonors at the same time, they say, both the country that he represents and the one to whom he is accredited. M. Flood, it appears, passes his days and nights in the most ignoble debauchery. One meets him, every day, in the streets in a state of complete drunkenness, and it is said that since he arrived in Texas, no one can boast of having seen him otherwise. However undiplomatic such habits may be, they would not perhaps have aroused such a degree of general indignation, in a country where unhappily they are only too frequent, if M. Flood had not made the mistake of boasting about them on every occasion as if they were a cause for admiration and friendliness with the Texians, and if he did not take care to repeat ceaselessly that it was in order to make himself popular among them that he resolved to be intoxicated from morning to night. My relations with such a colleague do not promise anything agreeable. So far they have been limited to a simple exchange of visits and I regret having to say that the only two times that I have seen M. Flood, he was dead drunk. M. Labranche is still at Houston awaiting him with impatience in order to hand over the archives of the Legation.

Galveston, June 17, 1840

Among the growing cities of Texas, which in the last years have made the most amazing progress, Galveston must be placed in the front line. In conversing with M. le Comte Molé, in my dispatch of April 17, 1839, no. 4, on the rapid growth that this city had already made at that time, I observed to him, however, that these bold and enterprising inhabitants would have many obstacles to combat and, what is harder perhaps, many prejudices to overcome to realize their project of making this port the main port of Texas; that (without speaking of all the inconveniences resulting from their insular position), the memory of the floods which had desolated them on several occasions and the fear of those that could at any moment swallow the city and all its population seemed to me must give rise to almost insurmountable obstacles; and that one must expect to see some company start up a rival establishment on one of the points of mainland in Galveston Bay. But the most distinctive trait perhaps of the American character is a spirit of imperturbable consistency, an obstinacy, an invincible tenacity in their struggles with material nature: once engaged in an enterprise, one can be assured that they will move heaven and earth to bring it to a good conclusion and that they will not give it up except in the last extremity. Towards the month of last February, several speculators joined together to buy on Point Bolivar, in the northeast of Galveston Island and in a fairly good situation, a vast site where they announced the intention of building a new city. Immediately the principal proprietors of Galveston and, at their head, the *Galveston City Company*, frightened by this formidable competition, went to the purchasers of Point Bolivar and proposed to them a considerable bounty if they would give up their purchase. The purchasers, who perhaps had no other aim, accepted eagerly and received a large part of the agreed price in shares of the city of Galveston. From this time the improvements have gone forward at an incredible rate. In the short space of time which has elapsed since my last trip here, entire new sections have gone up as if by magic; five spacious, convenient, and safe wharves have been built to load and unload vessels; a fine hotel, able to house more than 200 people, has been built where there was only a foul swamp last year; large trenches have been dug for the drainage of the swampy part of the town; sewers destined for the overflow of rainwater have been commenced; and there is talk of raising an immense dyke to protect the town from the ravages of the sea. Thanks to this extraordinary vigor and to all these improvements, the town of Galveston, tranquil henceforth on the Point Bolivar side, which was for it the most redoubtable, can flatter itself on having arrived at the realization of its dreams of maritime and commercial importance. What is most astonishing in these vast works, conceived, begun and finished in such a short time, is that they are made without money, and truthfully one does not know how. Of all the towns in Texas, Galveston, in spite of or rather perhaps even because of the importance of its trade, is the one which most feels the rebound of the financial crises in the United States and the disastrous effects of the depreciation of the only means of exchange which Texas yet possesses, Government vouchers. But the inhabitants appear to understand admirably the system of credit and exploit all its resources with a rare ability and marvelous good fortune. Another cause has contributed to the prosperity of this town. Since the seat of Government was transferred to Austin, Houston, in spite of its position at the head of Buffalo Bayou, has lost almost all its importance and more than half its population has come to swell that of Galveston which does not have today less than 3000 inhabitants. Doubtless this is very few for the commercial capital of a country, but one must realize that in the place where this town now stands there did not exist even at the end of 1837 a single house.

Your Excellency will not be less surprised than I was myself to learn that there are about 200 Frenchmen in the population of Galveston, and I feel a profound sentiment of satisfaction and pride in telling you that, according to the information furnished me by the authorities, they are all outstanding because of their excellent conduct, their love of order and their industrious activity. Several of them are at the head of respectable businesses. There is talk of creating a company entirely composed of Frenchmen in the militia of the town which will carry the name of *French Company*.

Tomorrow I propose to visit a new settlement started two months ago on a little island situated in the west pass of Galveston Bay. This little village, which has been given the name of Saint Louis, is composed of some twenty houses and 60 to 80 inhabitants, of whom a third, it seems, are French. Although its situation offers, in certain ways, undeniable advantages, its founders will have trouble, I believe, in making of it a formidable rival of Galveston, which is too far ahead today, and where too many interests are involved to fear much such competition.

Galveston, 26 June 1840

The inhabitants of Galveston named a committee to write me, the 20th of this month, a very flattering letter inviting me to a great public dinner. Without speaking of the boredom and the thousand annoyances inseparable from such solemnities, I would have liked because of my health, which is very sickly, to be able to withdraw myself from this honor. But I had already declined a similar invitation on the part of the inhabitants of this city three times before, and I feared lest a fourth refusal might offend them. Everything that took place couldn't have been better. The President, the Secretary of State, and all the city dignitaries were there. Your Excellency must think that no *toasts* were omitted. The first, offered by Colonel Love, chairman of the Banquet, to the King of the French, was received by thunderous applause which was prolonged repeatedly for more than fifteen minutes. When calm was reestablished, I rose to address the assembly some words of thanks which were well received and which I ended by proposing the health of the President of Texas. Then one drank to the President of the United States, then to the prosperity, to the glory of 'la Belle France', to the everlasting alliance between France and Texas, etc., etc., etc. When the list of *regular toasts* was exhausted, Colonel Love rose. "Gentlemen," he said, "I give you the health not of a prince or a potentate, nor an illustrious conqueror or a great statesman: I propose a toast to her who is the glory of her sex, to her who offers our wives, our sisters, our daughters a model perfect in every virtue; to the best of wives, to the best of mothers, to Marie Amelie of France!" At these words, the whole assembly showed an emotion, an enthusiasm that I would try in vain to describe. At these transports of love and respect which burst forth from all parts of the hall, I forgot for a moment that I was nearly 3,000 leagues from France; I believed myself once more in the midst of that people in whom the continuous sight of such high virtue, the example of which descends each day from the throne, has infused love and veneration for Our August Sovereign.

My modesty would not permit me to mention the toasts made several times to the chargé d'affaires of the King, if I were able to see in these overwhelming courtesies anything other than new proofs of the respect and friendship directed to France and its government.

M. Labranche, who arrived from Houston two days before, had been invited to the banquet. But he excused himself on the pretext that he would no longer be in Galveston on the appointed day. And yet he did not leave this city until the day afterwards. His suspicious irascibility was offended, I think, by the testimonials of esteem and affection of which I was the object at the time when they were letting him leave without any similar manifestations towards him. It appeared that similar reflections had at first decided his successor, M. Flood, to answer with some banal excuse the invitation that was also tendered him. Then he changed his mind, doubtless for fear of letting it seem too clearly the paltry feeling of envy . . . . He therefore came to the dinner and presented himself there in his usual condition, that is, completely drunk. A discourse that he pronounced, in a drunken and unintelligible voice, in response to the toast offered the President of the United States and that he persisted in prolonging during three quarters of an hour, in spite of the efforts of his neighbors to persuade him to hold his tongue, provoked at different times the laughter and jeers of the assembly. M. Lipscomb appeared very upset by the sad role played on this occasion by my colleague. In talking to me yesterday, he did not try to hide from me the deep discontent which the sending of such a representative had caused the Texian Government. "We must suppose," he told me, "that the President of the United States did not know M. Flood's habits when he chose him to replace M. Labranche; otherwise we should have to consider this choice as an insult to the People and the Government of Texas." "Indeed," he concluded by saying, "we cannot endure for long such a scandal, and I am going to write our Representative in Washington to request the recall of M. Flood."

Although the president and the secretary of state intend to prolong their stay in Galveston and I will not now find at Austin any members of the government, I start again tomorrow on the road to the capital. I have not been able yet to find other lodging than a sort of miserable wood cabin composed of three rooms for which I pay the *small sum* of 500 francs a month! I would like to try to set myself up in a manner a little more comfortable and above all more suitable, and I fear I will not succeed except by having a house built.

La Grange, July 17, 1840

Despite my desire to go to Austin as soon as possible in order to get a little rest there which my health greatly needs, I had to decide, in order to avoid the heat which has been excessive for two months, to travel very slowly. I even stopped several days at the home of one of my friends, Mr. Groce, who owns a plantation on the Brazos River where he has often invited me to visit him. Mr. Groce uses 180 Negroes on this magnificent plantation, which is second to none of the most beautiful properties on the banks of the Mississippi. He believes that, according to the least exaggerated evaluations, his cotton crop will reach 3,500 bales this year, which in calculating the price of the bale at 40 dollars, a price to which cotton has never yet descended, will yield the enormous sum of 140,000 dollars. In addition, and without speaking of the grain necessary for the consumption of the Negroes and his family, Mr. Groce harvested 20,000 bushels of corn last month, whose price is on the spot about $2.25 a bushel, and he counts on making about 100 hogsheads of sugar. These details can give Your Excellency an idea of the fertility of the Brazos lands.

But independently of the cultivation of tobacco, Texas possesses the elements of another trade which must one day prove even more profitable. I am speaking of the cattle trade. One cannot conceive in Europe of the immense hordes of wild cattle which crowd the prairies without end in the West, and whose number I would almost dare to compare to the grains of sands of the sea. They multiply with unbelievable rapidity, in spite of the continual pursuit of them carried on by the Mexicans and the Indians friendly to Texas or by the Texians who are bold enough to adventure into these lands. Not a week passes without thousands of these animals being brought to San Antonio, where they are sold to planters of the Colorado, the Brazos and other rivers in exchange for some products of the factories of Europe or the United States, of poor cotton fabrics, some boxes of ribbons, etc., etc.

Some attack by the Comanches is expected soon. For a month hardly a night passes without the inhabitants on the shores of the Brazos, and especially of the Colorado, having a part of their horses and mules taken off by these savages, despite precautions that they take to keep them safe. In vain numerous detachments of volunteers search the countryside in every direction; they find none of them. The Comanches, divided into bands of only two or three, travel through the forests by paths known only to them and where no one can follow them, keeping themselves hidden during the day in the neighborhood of farms and plantations, in impenetrable retreat, and then in the middle of the night, coming out to commit their thefts with an audacity and skill that defies all vigilance. The care that they take now only to steal horses and mules, to refrain, contrary to their habit, from all acts of violence against people, and not to disturb any of the numerous travelers, who circulate on all the roads in this season, is truly remarkable. From this it is concluded that the Indians, whose presence has recently been noticed by thefts in places fairly close to the coast where they have not yet dared to venture, fear to cause in the population too keen an alarm and that their intention is to concentrate at a single point and to attempt some bold move perhaps together with the Mexicans, who never cease to arouse them against Texas.

Austin, 26th July 1840

After an absence of several months, here I am again in Austin. I have found only two members of the administration here, the Vice-President and the Secretary of War. All the others are running around the country either for their pleasure or for their personal affairs. Also, at the end of the session, more than two-thirds of the population hurried to leave this city which is almost deserted today. It has been feared several times, especially in the month of last May, that the Mexicans and the Comanches would profit from the absence of the larger part of the population to attempt a strike against this capital which was unable to offer serious resistance. The alarms set forth each day on this subject in the newspapers, that do not cease reproaching the government for leaving thus the women and children to the mercy of the implacable enemies of Texas, determined the public to have certain fortifications built around the capitol where the meetings of Congress are held, in order that in the case of danger one can put the archives of the state there in safety, and if need be, even find a refuge against a superior enemy. But no sooner had work begun, than the opposition, whose outcry had provoked this measure, attacked with the greatest violence the system of fortification adopted, whose gigantic proportions according to them were completely useless against the feeble enemies Texas had to fear, and they took up again with great force their denunciation of the extravagance of the administration which seemed, to hear them, to have no other aim than the ruin of the country. In order to calm them, the president was forced to change completely his first plan and to adopt one more in conformity with the economical views of the opposition and the financial state of the country.

One was therefore limited to surrounding the capitol with a ditch of several feet and with a simple palisade which, to tell the truth, seems to me, with people like Texans to defend it, to be sufficient against enemies like the Comanches and the Mexicans. But the opposition was not beaten. According to them, these pretended fortifications, good only to amuse children, would not be able to stop for an hour a band of fifty Indians, and jibes rained from all sides on . . . the "admirable fortifications" that are known no longer except by the name of "*Lamars's folly*."

In spite of the care that I took coming from Houston to travel slowly and to avoid the great heat of the day, I already have to deplore the terrible effects of the burning sun of this country. A black servant that I bought in New Orleans died yesterday after an illness of 36 hours, caused by the fatigues of the trip. Two other of my people are very dangerously ill, and I myself am very sick.

The condition of the President does not seem to improve. His friends begin to worry seriously about him and doubt that he will be able to come back to his residence here before a long time.

Nothing new yet on the subject of the Indians except that the detachments sent to pursue them did not succeed in discovering them although they continue at a multitude of different points to steal mules and horses. Some travelers recently arrived from Matagorda claim to have met a little above Victoria, on the banks of the Guadalupe, a dozen of these savages divided into two or three bands, who took flight on seeing them. As for several years the Comanche have never dared advance so near the coast, little faith is placed in this report.

Austin, 1st August, 1840

Your Excellency knows how General Samuel Houston, whose unquestioned services to the Country would seem to have assured him an indestructible influence, happened, soon after his election to the Presidency, to have raised against him violent opposition on account of personal misconduct and his scandalous habits of debauchery, not less than by the arbitrary conduct, the disorder and despotism of his administration. This was a misfortune for Texas, for if General Houston had wished to make a better use of the precious faculties with which he was so amply endowed by nature, no one would have been able to do more for the prosperity of the young State to whose formation he contributed so much.

General Lamar arrived in Texas with a high reputation of ability: his integrity was recognized by all, and he had given many proofs of his devotion to the Country and had made himself noted for his courage on several occasions, notably at the battle of San Jacinto whose success his friends even today attribute solely to the coolness and decision that he showed there. His opposition to the despotic measures of General Houston and above all to the plans for annexation to the United States won him many supporters. He was elected by an immense majority and his administration commenced under the happiest auspices. From all sides the most outlandish and ridiculous praises were showered on him: as a statesman, he was placed in the same rank as the greatest ministers of France and England; modestly his eloquence was compared to that of the powerful orator of our revolution whose name he bore; and his military glory must one day equal that of the Great Captain, his other namesake. Finally, General *Mirabeau Bonaparte Lamar* was regarded as a providential man, sent on earth to establish on an unshakable foundation the greatness and prosperity of Texas. Hardly two or three newspapers, remaining faithful to General Samuel Houston and his hatred of his successor, dared to venture some timid protestation against this anticipated apotheosis and to mingle their discordant voices with this universal hosannah! But alas! Such touching agreement could not last; great popularity scarcely lives more than a day, above all in Republics. The People, and absolute people even more than others, changeable in their caprices, terrible in their reversions, never cease to make their passing favors dearly bought, and General Lamar must learn at his expense, that, as it has been repeated so many times, there is only one step from the Capitol to the Tarpeian Rock.

Among those who supported the candidacy of General Lamar with the greatest warmth, were found the former friends of General Houston that personal grievances, the resentment of wounded pride and betrayed ambition, and, above all, the hope of obtaining the favor of a new President had thrown into the party of the latter. When General Lamar, who could judge how much he could count on them by their conduct to their former patron, arrived at the Presidency and did not show himself more disposed than his predecessor to do justice to their pretentions, they turned against him with fury. At first they attacked his personal character; not being able to attack him on the score of his probity, they struck out against the disorder and profligacy of his morals; they accused him of living publicly with colored mistresses who are his slaves and by whom they pretend, in fact, that he had several children; and though he thus would make himself guilty of a sin very common in all the Southern States, they thundered with *righteous* indignation against the Head of the Republic who would not fear causing such a scandal.

Austin, 25 October 1840

Your Excellency will have been informed by M. Desages of the cruel illness to which I almost succumbed.

After having been forced to keep my bed nearly three months, I now find myself almost entirely recovered, and I hope to be in condition to get back to work in a few days.

I recently addressed to Your Excellency duplicates of my [dispatches] no. 15, 16, 17 and 18. At the time of my departure from Galveston I had entrusted these dispatches to a person of my acquaintance who was going to New York and who, according to the details published soon afterwards in the newspapers, was massacred by the Indians near Tallahassee in Florida.

As the mail service has been even more irregular than usual during the great heat of the summer, I have thought it wise to send you also duplicates of my nos. 19 and 20.

Today I profit from a safe opportunity to send you a dispatch of the date of August 1st, no. 21, that I was preparing to send when I was stricken ill.

Austin, 26 October 1840

It [the opposition] has been careful above all not to say a single word about its true and only complaint against the President. The part taken by General Lamar in the legislative action that transferred the seat of Government to Austin, his well known perference for the western section, his efforts to bring immigration there, such is in reality the cause of the violent diatribes of which he has been the object. This is what has aroused the population of the East and North-East against him, for those who set themselves up as his political adversaries belong, almost without exception, to these two sections of the country. It is so true that at the bottom of all these quarrels, as I have told Your Excellency, there are only the petty interests of locality, that it is on this basis that the question was everywhere clearly and solely placed at the time of the last elections, at the beginning of September, and that the opposition, despite its efforts, has not been able to give a political character to it.

Austin, 30 October 1840

Your Excellency recalls that, in the course of

the month of July, I had the honor of informing you of circumstances that seemed to indicate the presence of Indians at several places in Texas and their intention of concentrating secretly in order to attempt some attack with all their combined forces. The measures taken by the government to prepare for this danger must have been useless, for these savages succeeded in hiding their course from all eyes, and no one suspected them of enough resolution, enough audacity to cross the entire country and come to attack the Texians as far as the Gulf of Mexico.

In the morning of the 10th of August, a messenger brought the news here that a troop of Comanches and Cherokees to whom were joined a certain number of Mexicans had attacked the town of Victoria, on the Guadalupe River on the evening of the 7th; that after having massacred the greater part of the population and pillaged and burnt the town, these savages fell upon Linnville, a new settlement placed at the end of Matagorda Bay; that they destroyed everything there, killed all the inhabitants to the last one; and that they then spread out into the neighboring country, putting everything to fire and sword.

One cannot imagine the consternation caused by the first news of these sad events, that fear enlarged more, as always happens. People were positive that the number of savages was no less than 5 to 6,000 and that one should expect, at any instant, to see them descend on the city of Austin, which they would ravage as they had Victoria and Linnville. There was even a rumor in Galveston, which was repeated in several newspapers of this country and then in those of the United States, that the Capital of Texas had been reduced to ashes by the Comanches and the population killed or led into slavery.

The Government, without allowing itself to be intimidated by these reports, evidently exaggerated and arriving from all sides, gave General Felix Huston, Major General of the Militia, which had just arrived at Austin, the order to march against the Indians with as many volunteers as possible. On the 11th at daybreak General Huston set out at the head of 170 men, resolved in spite of their numerical weakness to hunt out boldly the Comanches and cut off their retreat.

The following morning at 6 o'clock he was camped at *Plum-creek* about 6 miles from Austin, when his spies came to warn him that the Comanches were approaching. Hardly had he time to draw into battle line his little troop, grown by recruits made on the way to 220 men, before the Indians appeared. The number of warriors, without speaking of women, children and old men following them, amounted to 8 or 900. With their rear guard were more than 500 horses or mules laden with loot. This extreme difference in forces did not frighten the Texians who bravely barred the passage of the Comanches. The battle began immediately with a great desperation on both sides; but in less than an hour the savages, beaten on every side, took flight and fled into the mountains, leaving some sixty dead on the battlefield, among them their principal chiefs, and 7 or 8 wounded, and abandoning to the conquerors all their loot and more than 700 horses. Five or six prisoners that they had brought from Victoria and Linnville were successfully snatched from their hands. But these unfortunates, that the savages had the cruelty to riddle with arrows and spears at the beginning of the battle, died in several hours from their wounds, with the exception of a young woman married only two days before the arrival of the Indians to the Collector of Customs at Linnville and whose husband had been killed by them.

General Huston, whose men were weakened by fatigue, had to give up the pursuit of the enemy into the mountains, and after having uselessly searched the neighboring country for two or three days in the hope of surprising some other party of Comanches, he returned here having had only one man killed and three wounded on the expedition.

Before his return there was time to receive correct information on the events at Victoria and at Linnville. In the first of these towns, which has a population of 8 or 9,000 souls, the Indians, after having sacked two or three isolated houses and having killed four or five people, withdrew before a hundred volunteers assembled in haste to repulse them. At Linnville, which was composed of 5 or 6 houses only and the stores of the Custom House, where at this time there was about 150,000 dollars worth of merchandise, they had stolen or destroyed everything and killed three of the inhabitants as well as some Negro servants. However regrettable these misfortunes were, one must be pleased that there were not greater ones to deplore: they were, praise God, much less than the terrifying reports that were first circulated.

I would not be able to pass over in silence an overture on the part of the Texian government of which I was the object . . . When the news of the attack on Victoria and Linnville by the savages arrived here, I was in my bed, almost dying. The secretary of state, fearing that the exaggerated reports which circulated publicly might have caused me great uneasiness, came quickly to reassure me. He told me that it was completely unlikely that the Comanches would dare to attempt an attack on Austin; that in any case, before arriving here, they would meet on their passage the troops of General Huston; and that besides there remained in the city more than 300 armed men, which was more than sufficient to repulse them. Finally he insisted that I should be moved to the Capitol whose wall had just been fortified, as Your Excellency knows, and where he told me lodgings had been prepared for me. I thanked M. Lipscomb, and as I was also convinced that the Indians would not dare to appear before Austin and besides I could not leave my bed, in the state that I was, without great danger, I did not accept his proposition, and I satisfied myself, in order to reassure my people who were very frightened, by having arms distributed to them and setting up a guard of four men in my yard.

The President has been so ill for several days that it was feared at one moment that only a few hours of life remained for him to live. His condition is a little better since, but it still causes deep concern to his friends.

Austin, 6 November 1840

As I had announced my intention to Your Excellency, not being able to find suitable lodging, I have made up my mind to have a house built on a very beautiful location that I have bought for this purpose. Unhappily the difficulty of procuring the necessary materials and, above all, my long illness have greatly retarded the work, and I fear I will not enter the new house before springtime. I am thus resigned to entertaining in the humble dwelling where I am encamped rather than installed and to do the honors as best I can. I had all the cabinet and several senators to dinner there, several days ago. I also invited the President, and he did me the honor of accepting. However, the state of his health did not permit him to keep his promise, and he had Mr. Burnet express his regrets to me. I intend, during the session, to entertain at my house twice a week, either at dinner or in the evening, the members of Congress and the most important people of the Country. People are very moved here by these invitations, by these courtesies which are, in this country more than anywhere else, powerful means of influence and an indispensable condition of success.

Austin, 14 November 1840

The two expeditions sent against the Indians under the command of Major Howard and Colonel Moore have succeeded beyond all hopes. The former having met, about 80 miles above San Antonio, a band of savages, which included more than 300 warriors, attacked them despite the extreme inferiority of his forces and completely routed them after having killed nearly half of their number.

Colonel Moore was not less happy on his side. Guided by a detachment of Lipans, he unexpectedly fell on a camp of Comanches placed near the source of the Colorado, made a terrible slaughter of them and went away with about sixty women and children. He just arrived here with his prisoners of whom several have already succeeded in escaping from his hands; and the inhabitants of Austin, as a testimonial of their gratitude for the signal service he has rendered the Country, hastened to offer him a public dinner.

One awaits also from day to day the chargé d'affaires of the United States, who, since his arrival in Texas, has not yet visited the Capital. Mr. Lipscomb, who continues to express himself publicly with a fairly severe frankness on the attitude and undiplomatic habits of Mr. Flood, told me, some time ago, that he wrote him to tell him that it would be fitting if he at least put in an appearance appearance at the seat of Government, and to let him clearly see that if he did not comply with this invitation, the cabinet at Austin would feel obliged to address some observations on this subject to the cabinet at Washington. This desire of the Texian administration to see the Foreign Ministers reside, at least during the session, at the Seat of Government is only very natural, but yet the principal officers of this Republic, and Mr. Lipscomb above all, should commence by setting the example.

Austin, 25 November 1840

General Lamar is still resolved to leave as soon as his health will permit him, and it is said that he will address Congress a message on this subject tomorrow or the day afterwards. I will the more regret seeing General Lamar leave the control of affairs, because Judge Burnet, who will be called to replace him, cannot be counted among the number of friends of France.

General Samuel Houston has been here several days, and although very ill, he has been from the day after his arrival occupying his seat in the Chamber where he will exercise, I believe, a great influence. I have just learned this instant that the chargé d'affaires of the United States, Mr. Flood, who had not yet appeared in Austin, has just arrived in town.

Austin, December 5th 1840

For nearly a month the rain has not ceased to fall in torrents in all parts of the country situated below Austin. The Brazos, the Colorado and all the streams and creeks tributary to these two rivers are flooded: the country is nothing but a vast plain of water on which float inhabitants and stock carried away by the violence of the destructive elements. However great were the losses caused by this terrible scourge, they are nothing, however, compared to those that one would have had to deplore if the season had been more advanced. One of the great inconveniences which resulted from it today is the complete interruption of all communications and the absence of news from the places desolated by the flood, which contributes still more to augmenting the general anxiety. The contractors charged with the transport of the mail have declared to the government that it is impossible for them to continue service. Here we are thus threatened to remain a long time still perhaps without mail from Galveston. Two sacks that I had sent more than three weeks ago with my dispatches for Your Excellency were returned to me yesterday from Bastrop, 35 miles from here, where they were necessarily detained. I do not know what means to employ to send them again.

The citizens of Austin decided some time ago to give a ball for me as a testimonial of their esteem. M. Flood having arrived meanwhile, it has been decided very suitably and to my great satisfaction that the ball will be given for the chargés d'affaires of France and the United States. This party will take place tomorrow.

Austin, 14 December 1840

Following a severe attack which kept him for two days in extreme danger, General Lamar, acceding to the orders of his doctors and to the pleas of his friends, decided to leave. The 11th of this month, he addressed Congress to ask authorization to go to the United States, which was immediately granted him, and yesterday morning he left the President's House. It is feared that his condition will not let him support the fatigues of travel and that he may be forced to stop on the way.

The last weeks that General Lamar passed at the seat of Government went by in sad solitude. Since he has had to give up entirely conducting affairs of state and since all hope of recovery has disappeared for him, the crowd of functionaries of all sort who habitually surrounded him, his usual courtiers, and his most devoted flatterers, have left him to pay their respects and their protestations of devotion to his successor, Judge Burnet; and when I went, the evening before last, to take leave of the President, I scarcely met two or three friends who remained faithful around his bed. The departure of this first officer of the Republic, leaving the capital probably never to see it again, caused no more excitement than if it were a question of a simple private citizen. The official newspaper itself, in the middle of a pompous panegyric of Mr. Burnet in its last number, only found a few expressions of regret, very cold and dry for one who, during his stay in power, never ceased being the object of the most insipid and exaggerated adulation, for the man who ( his enemies even could not refuse him this justice) has always fulfilled, if not with skill and good humor, at least with conscience and integrity, the difficult duties which were imposed on him, and who, in several circumstances, has rendered his country unquestioned service. A striking example, among others, of the inconstancy of the multitude, which should serve as precept to those whom an unhappy and blind ambition ceaselessly pushes to pursue a vain popularity at the price of the most painful sacrifices! General Lamar, when setting out on his journey, had the extreme consideration of having his carriage stop at the door of my house to be able, he told me with an emotion I could not help sharing, to shake my hand a last time.

A numerous party of Comanches, surprised fifteen days ago by Major Howard near San Antonio, has been almost entirely destroyed. It is certain that the savages lost more than 80 of their warriors in this affair, without the Texians having a single man killed.

Austin, 24 December 1840

Congress has just declared again against the idea of changing the seat of government. A proposition was introduced in Congress that there should be, during April 1841, a census taken of the population of the Republic. This operation would have had the result of adding, in the September elections, to the number of representatives from the Eastern counties where the population has greatly grown during the last three years and would have assured them of a majority in the next Congress which they would not have failed to profit from in order to move the Capital elsewhere. Warmly defended by all the delegation from the East, this proposition was attacked with no less vigor by the delegation from the West, and after two days of debate, the chamber, by a decision which is in the interest of the Country, decided against it by a majority of a single vote, 18 members having voted for it and 19 against.

One of the last discussions of this assembly had a bad reaction from the public and almost led to even more regrettable consequences. It was a question of a supplement to the salary claimed by Mr. Burnet, as charged in the interim with the functions of the President. General Houston opposed this demand in a very animated speech in which, discarding the moderation and the courtesy of language of which the most irritating questions, the most churlish personalities, ordinarily cannot deprive him he attacked with extreme bitterness not only the public acts but the private character of Mr. Burnet and made clear and direct allusion to a disagreeable affair in which the latter was several years ago accused and, what is worse, convicted, it appears, of having stolen *two pigs* from one of his neighbors. Mr. Burnet, who in spite of his surly and annoying disposition, has never felt the least inclination for the profession of arms, had at first resolved, wisely no doubt, not to provoke an encounter and to answer only with silence and disdain an insult aimed too low, he said, to reach him. But the pleas, the threats even, of those of his friends who told him that he would disgrace himself if he did not demand redress for such a bloody outrage, forced him to send to General Houston a challenge which the Secretary of War, Dr. Archer, took the responsibility of delivering to the latter: General Houston refused to answer Mr. Burnet's letter. He only said to Mr. Archer that everyone knew that several years ago, because of an unhappy event ( a duel in which he had the misfortune to kill his adversary), he had resolved never to send or receive a challenge, but even if his scruples in this respect were not invincible, the respect that he owed himself would not permit him, in any case, to fight with a man whom he had been obliged to point out and to brand publicly as a thief. As a result of this conversation, Mr. Archer published in the official newspaper two letters in which he addressed the most insulting provocations and the grossest abuse to General Houston. The latter did not deign to reply and only spoke of it with a smile of pity. "Have you seen," he asked me the other day, "Archer's magnificent lucubrations? How does his style seem to you? What verve! What elegance! Above all, what good taste! In truth this dear doctor makes me unhappy. He is a good man, whom I have always liked, and I pity him with all my heart. So much for the influence of bad company! I have predicted that those people will drive him mad, if they don't succeed in making him a dishonest man." These discussions, which have produced a painful impression on the public, happily appear to have ended, and it is to be hoped that they will not start again. It is agreed that General Houston did the first wrong; but people blame no less severely the conduct of Judge Burnet and his friends whom this affair has hurt in the opinion of the reasonable people in their own party.

Austin, 19 January 1841

The [Catholic] Church, which will be placed under the protection of Saint Mary in memory of the august and virtuous Princess whose loss cost France so many tears, will be built on a plot of 5 acres, of which I made a gift for this purpose to the Catholic clergy, beside the place where I am at this time having my house built.

Austin, February 19, 1841

The President has finally found a Secretary of State . . . . It was impossible for M. Burnet to make a more unhappy choice in every respect. M. Mayfield, in spite of his ambitious pretensions and the air of importance that he gives himself, is of a less than mediocre capacity and of an extreme ignorance. Without speaking of his private character, which is not very respectable, of his gross and brutal manners, of his surly and trouble-making disposition which makes him many enemies; his lack of honor and loyalty towards his own party during the last session and his indecency in violating the most sacred engagements have brought him into disrepute as a man of politics and have drawn to him the hatred and disdain of the friends of the government as well as of the opposition.

General Houston's happy reformation of habits since his marriage, and the dignity with which he was able to conduct himself, despite several mistakes, during the last session, have made him regain all his popularity and he is now certain of being elected President next September by an immense majority . . . .

12 May 1841

As for what concerns me personally, M. le Ministre, if I listened only to my desires or my personal interests, I would pray Your Excellency to be willing to grant me the leave that I had the honor of asking of you last February and to permit me to return immediately to France to which I am urgently called by the need to restore my health and to tend to my personal business. But as the interests of the service supercede every other consideration, I must say frankly to Your Excellency that my departure, at the point where things are now, would have, in my opinion, great drawbacks.

Mandeville (Louisiana), 14 August 1841

I have also received a letter from General Houston. He appears not to have the least doubt of his election, for he begs me to have him made in Paris a *uniform* (court-dress) that he wishes to wear the day of his inauguration. I had the occasion, in my correspondence with the Department in 1839, to make allusion to the outlandish taste in dress that General Houston brought back from his trip among the Savages. He has shown himself faithful to this taste in the choice of the uniform that he asked me to have sent him. This costume is to be composed of a complete suit of green velvet all embroidered in gold; of a little Spanish-type cape also of green velvet and covered with rich embroidery; and in addition a hat à la Henry V, hidden by an immense plume in three colors. It is in this strange outfit that the future Head of the Republic of Texas intends to take his seat in the Presidential armchair. I am hastening to write to Paris to carry out the instructions of General Houston.

# THE PIG WAR

Besides his connection with the French Legation building, Alphonse de Saligny is most often remembered as a participant in the so called Pig War. The following excerpts from the French Foreign Office microfilm in the Austin Public Library give a fuller account of this altercation between Saligny and the innkeeper Bullock that became an international incident.

From Saligny to the Secretary of State, J. S. Mayfield

Austin, Feby 19th, 1841

To the Hon. Secretary of State - Austin

Sir:

It is with profound regret that I find myself forced to call your attention to facts of an extremely grave nature of which this city was this morning the theatre.

One of the people belonging to my house, Mr. Eugene Pluyette, passing in the street, was without any provocation on his part, assaulted by a man named Bullock, a resident of this city, who having first assailed him by throwing himself upon him with a stick, at the same time made use of all kinds of menaces and atrocious imprecations. My domestic after having repulsed successfully the attacks of Bullock, wished with a moderation worthy of praise, to pursue his way, but this miserable man, without the *sangfroid* and presence of mind of my domestic, undertook to follow him, and the affray was commenced again in a more serious manner.

These acts, Sir, constitute one of the most scandalous and outrageous violations of the Laws of Nations, and they assume a much more serious aspect when it is reflected that they are but the realization of menaces made long time in advance and that even (as I have but this moment learned) they have already, at two different times been preceded by facts (or occurences) of the same nature.

I think I ought to add for your information that Mr. Eugene Pluyette who has been for a long time in my service has always been remarkable not only for his amiable and inoffensive character, but for his unimpeachable probity. It appears on the contrary that this man called Bullock is very far, according to several reports, from enjoying a spotless reputation.

A. de Saligny

From Richard Bullock to President David G. Burnet

City of Austin, 20th Feb., 1841

To His Excellency, David G. Burnet
President of the Republic of Texas

Your memorialist would most respectfully represent to your Excellency, that in pursuance of his lawful business which, since he resided in Austin, has been to accomodate with board travelers and other persons - that the Chargé d'affaires of France, M. Saligny, has become his debtor, to the amount of two hundred and seventy dollars and seventy-five cents (27075) par money . . .

Your memorialist further states, that soon after his debtor refused payment (for the first time) that he has suffered furthermore detriment in the loss of hogs, which have been most maliciously and wantonly killed with pitchforks and pistols used by his debtor and a Frenchman in his employ called Eugene, or by servants under his debtor's directions. He supposes the number of hogs killed by them to be between 15 and 25, the value of which he thinks would be about one hundred dollars ($100) par money. Your memorialist shewith further that the fence of his debtor is not a lawful one, it being rarely ever kept up.

Your memorialist said above that his hogs were wantonly and maliciously killed, his reason for the declaration is, that, his debtor was not at all injured or incommoded, his hogs occasionally, as well as the hogs of other neighbors, only went under Saligny's horsetrough, fed for the most part in an almost open lot, there being no garden or kitchen that they either molested or disturbed.

Your memorialist is sure, that had another man, amenable to the laws of this Republic, done like acts, he would have recovered in a court of justice the value of his property, having witnesses abundant to prove the condition of his fence and pailings, as well as the killing of his stock.

Your memorialist respectfully asks of you, as the chief magistrate of our country, protection in his rights (as his debtor is not amenable to our laws) and he is constrained confidently to expect and he fully believes you will have justice done to one of your humblest citizens.

and your memorialist will ever as in duty
signed Richard Bullock

From J. S. Mayfield, Secretary of State to Saligny

Department of State
City of Austin
Feb. 20, 1841

Your note has been submitted to the President. He regrets exceedingly the occurrence alluded to in your note, and directs me to assure Mr. Saligny, Chargé d'affaires of France, that he is fully sensible, and awakened to the vital importance of maintaining in this respect most scrupulously and inviolably the laws and usages of Nations; and that the Government of the Republic of Texas will at all times exercise to the limits her authority, to bring punishment to any person who so far forgets the honor and dignity of her institutions and law and the respect due those nations, when friendly relations are established, and who may have resident Ministers here, as to violate in any way, the long established rights, privileges, or exemptions belonging to public Ministers from abroad, or their family, suite or servants, or those who in any wise are entitled to participate in the inviolability attached to this public character.

From Saligny to the Secretary of State, Feb. 21, 1841

. . . It is a matter of extreme urgency in the interests of our respective governments that the attempt of the said Bullock has rendered him liable to receive a prompt and exemplary punishment. This individual, who has by no means *restrained* himself . . . to *provoking* Mr. Pluyette, but who has attacked him on three different days, sometimes by throwing stones, sometimes with a stick, and even with a hatchet, inflicting on him each time wounds that were more or less serious, has since yesterday, again used towards him horrible threats. In consequence of these I have therefore thought it proper to order my servants never to go out without being armed, and I have expressly recommended to him to make use of his arms to repulse any new outrage which Mr. Bullock may attempt to commit upon him.

The Secretary of State to Saligny, Feb. 22

. . . I beg to refer Mr. Saligny, Chargé d'affaires of France to the accompanying letter of Mr. Jewett the proper law officer, who has as will be seen in pursuance of instructions from the Department instituted the necessary legal measures for a judicial examination into the matter. From this it will be seen that testimony will be received tomorrow at 3 o'clock p.m. at the Senate Chamber, when and where it is expected Mr. Saligny will cause to be laid before the Hon. A. Hutchinson one of our District judges and a member of the Supreme Court, any evidence at his command touching the transaction, which is the subject of discussion . . .

Saligny to the Secretary of State, Feb. 22

. . . I would even consent in the case in which you have made the demand, that my domestics should be heard by simple declarations - but as to permitting that they should appear as witnesses before the Judicial authorities of the Country, particularly when it relates to a question in which the dignity of France is deeply concerned, is what I could not do without completely forgetting the obligations which are imposed on me, without a culpable abandonment of principles and of privileges, the preservation of which uninfringed, is one of my first duties.

A. de Saligny

Secretary of State to Saligny, Feb. 25

The subject matter of complaint having been referred by the Executive to the appropriate Department the judiciary, the Government has no further control in the matter, and the alleged offence could only be reached in the manner indicated . . .

Saligny to the Secretary of State, March 21

The *memoire* of Mr. Bullock can be nothing, in the eyes of every enlightened and impartial man, but a plea invented after the event (it is easily imagined for what purpose) and it would not be proper of me to refute one by one all the false allegations, all the fabricated and completely barbarous points that it contains. Thus I will limit myself to telling you that it is untrue that I ever refused to pay what I owe to Mr. Bullock; that far from that, I offered him, time and time again, by several persons, to pay him immediately not, it is true, according to the account drawn up by him and which everyone has termed scandalous (most scandalous imposition), but according to that fixed by arbitration, conforming to the *known* prices of Mr. Bullock.

But this individual has always refused, and during entire months, he has not ceased to vomit forth against me the grossest insults, to spread about from door to door the most outrageous calumnies.

As for the complaint of Mr. Bullock relating to his pigs, here is the truth. I have for a long time suffered, and I still suffer every day, like every one, from the numerous pigs which infest the city. Every morning one of my servants spends two hours repairing and re-nailing the boards of my fence that these animals break in order to come eat my horses' corn; 140 pounds of nails have been used for this purpose. One day three pigs came even into my bedroom; ate the linen there and destroyed some papers. Another time a dozen of these animals, in order to eat the corn, rushed into my stable and trampled one of my servants who was rescued half dead, with great difficulty. It was then that following the example of all my neighbors, I ordered my people to kill all the pigs that should come into my yard. But this did not apply specially to the pigs of Mr. Bullock, which did not carry on their backs the name of their master and that it was impossible consequently, to distinguish from the others. Following my orders, five or six, it seems, were killed *in my yard*, by my servants. Did they belong to Bullock or to someone else, I don't know.

For several months, Monsieur, the Texian Government, without paying attention to the friendly remarks that I contented myself to address to it; without appearing to appreciate as I have the right to expect, the sentiments that moved me to a patience, to an excessive forbearance, has permitted Bullock to vomit, everyday, against me, the Representative of France, the grossest insults, the most insulting calumnies.

Saligny to Secretary of State, March 25

France has just been insulted again in this city, in the most outrageous manner, and this time, it is on the person even of its Representative that the outrage has been committed. Yesterday evening, at the moment when I was getting ready to enter the yard of the dwelling of the chargé d'affaires of the United States, to whom I was going to make a visit, the inn-keeper Bullock, who had been following me for several minutes, with a decided and threatening air, that I could not help but notice, hurled himself at me and told me in an insolent tone that he forbad me to come into his house. I answered him calmly that I was going not to his house but to that of Colonel Flood. "That is not true, he said then; you are at my place and the first time that you come back here, I shall beat you up. You have been warned; henceforth I shall not limit myself to words but I shall act." While speaking thus, he shook his fist at me threateningly and acted as though he wished to hit me. I told him to mind what he was going to do. Then he took me first by the collar, then seized me by the arm violently. However, disconcerted by my *sang-froid*, he released me and I continued on my way without paying attention to the threats and menaces that he continued to utter.

You will understand, Monsieur, that I would not be able to remain any longer with a Government which, far from being able to let me enjoy the respect and protection due a Representative of a friendly Power, would not even have the power or the desire to protect my life from the assaults of a scoundrel. I beg you therefore to let me know, as soon as possible, the measures adopted by your Government to prevent Bullock from carrying out his threats.

Secretary of State to Saligny, March 29

I am aware that in matters of this kind, you assume the position that the constitution and municipal regulations of our nation are not to be invoked, when the rights of another are alleged to have been violated in the person of its Representative. It seems to me that it will require but little reflection on your part, to convince you that your views of this subject are incorrect. It is admitted that the law of nations alone is to be resorted to for the purpose of ascertaining the right and immunities of foreign ministers, or for determining what acts on the part of others constitute offences against them, or the Nations which they represent, but the offence once ascertained, where do you find the rule which provides for the mode or extent of the punishment? It is believed that none such is to be found in the works of any writer on international law, and consequently all nations are compelled to resort to their own municipal codes, for the means of carrying out and enforcing within their own territory the provisions of the laws of nations.

Saligny to the Secretary of State, March 31

Your note of the 29th was given to me yesterday. This note, which resembles more an impassioned counsel's address in favor of the said Bullock and a slanderous libel against the Chargé d'affaires of France, than a Diplomatic communication, is of such a nature, it is conceived in a spirit and written in a tone so injurious to me, that I shall abstain from replying to it . . . I will limit myself to acknowledging its reception to you; and, until I have received further orders from the Government of the King, to whom I have sent a copy, I think it necessary to suspend all relation with a Government which, when the Representative of France addresses it to obtain justice for bloody and repeated outrages, answers only by insulting words.

Secretary of State to Saligny, April 5

. . . I have the honor to state that you can obtain your passports when you choose to demand them; or can remain in this country, with the exercise of your ministerial functions suspended in that unofficial character you have thought proper to assume. In as much as you place yourself in the extraordinary attitude of declining the exercise of your functions as Chargé d'affaires of France, and all intercourse with this

Government; as a matter of course your immunities and priviledges as such cease, and no further protection can be claimed by you than what the laws extend to her own citizens.

Saligny to the French Foreign Minister, May 6 - from Houston

In spite of my desire to facilitate by every means in my power the adjustment of the difficulties caused by the strange conduct of the cabinet at Austin; in spite of the extreme moderation which I have not ceased for one minute to exhibit, it has been impossible for me to conclude anything.

To prolong any longer my stay in Austin would have been an inexcusable weakness towards the administration; besides, it would have been an act of imprudence; for Bullock, more audacious than ever, since he feels himself supported *officially*, announced openly the intention of killing me with his rifle the first time he should see me, and although very little frightened personally by the threats of that individual, I should not, by exposing myself wantonly to new outrages, risk further aggravating difficulties already so serious. I resolved therefore without delay to get ready to depart.

I am still willing to receive overtures that he [the President] might wish to make me; and if there is a way to arrive at some settlement, I would be delighted. In that case I would return soon to Austin, where I left a confidential man of business and a servant to take care of my establishment, or I might rather decide to pass some time at Galveston, where I intend to go tomorrow and where the President is expected next month. If, on the contrary, M. Mayfield (Secretary of State) persists in his blindness, then I will ask permission from Your Excellency to submit my views on the course to follow in this affair, which disagreeable as it may be, should not after all have very serious consequences, and that we can even, with a little skill and determination, make serve our interests.

---

*And on this note, Saligny left Austin - never to return. The Pig War was ended!*

*Typesetting by*
EKHOLM TYPESETTERS

*Printing by*
CAPITAL PRINTING COMPANY

*Paper supplied by*
LONE STAR PAPER COMPANY

*Binding by*
CUSTOM BOOKBINDERS

*Design by*
WILLIAM D. WITTLIFF